Do the Math!

$3x + x = 4x$

$\dfrac{4x^2}{2x} = 2x$

$3 + x = 4$
$x = 1$

$4(x + 1) = 4x + 4$

Volume VI
Introduction to Algebra
Applications of Linear Equations
in One Variable

Suzanne Bower

Copyright © 2018 Suzanne Bower

All rights reserved. No part of this publication may be reproduced, stored in a retrieval system, or transmitted in any form or by any means, electronic, mechanical, photocopying, recording, or otherwise, without the prior written permission of the publisher.

Contents

Chapter 1 - Translating Word Phrases
            into Algebraic Expressions   5

Chapter 2 - Number Problems, Age Problems,
            Geometry Problems   13

Chapter 3 – Money Problems, Mixture Problems,
            Interest Problems   23

Chapter 4 – Motion Problems   35

Review   42

Test   45

Answers to Exercises   48

Chapter I – Translating Words into Algebraic Expressions

In order to use algebra to solve problems, it is necessary to translate word phrases into algebraic expressions to set up an equation. Here are some examples.

The following algebraic expressions can be used to express the corresponding word phrases:

n + 5 The sum of a number and five
A number increased by five
Five added to a number
A number increased by five
A number plus five

n – 5 The difference of a number and five
A number decreased by five
Five less than a number
Five subtracted from a number
A number minus five

5n The product of a number and five
Five times a number
A number multiplied by five

2n Twice a number

$\dfrac{n}{2}$ The quotient of a number and two
A number divided by two

$\dfrac{n}{2}$ or $\dfrac{1}{2}n$ Half a number

Examples: Write the following as algebraic expressions.

1. Five more than the product of a number and six

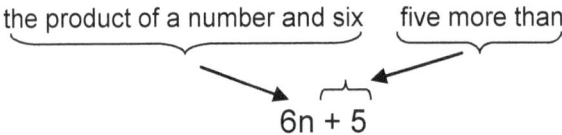

2. The product of five more than a number and six

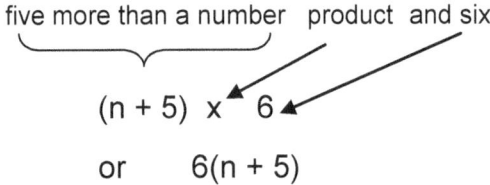

3. The sum of five times a number and eight

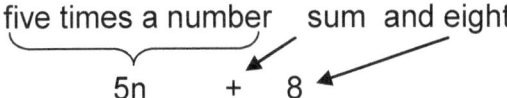

Practice problems: Write the following as algebraic expressions.
1. Six more than the product of a number and eight.
2. The product of three more than a number and seven
3. The sum of nine times a number and eleven

Example 4: Write algebraic expressions using one variable for the number of male students and the number of female students in a class of 35

Solution:
n = the number of male students.
$35 - n$ = the number of female students

Practice problem 4: An animal shelter contains 56 cats and dogs. Write algebraic expressions expressing the number of each.

Example 5: Write algebraic expressions using one variable to express two consecutive integers.

Solution:

$$n = \text{the first integer}$$
$$n + 1 = \text{the second integer.}$$

Practice problem 5: Write algebraic expressions using one variable to express three consecutive integers.

Example 6: Write algebraic expressions using one variable to express two consecutive odd integers.

Solution:

$$n = \text{first even integer.}$$
$$n + 2 = \text{the second integer}$$

Practice problem 6: Write algebraic expressions using one variable to express three consecutive even integers.

Example 7: Write an algebraic expression for the price of a television with a mark-up of 15%.

Solution:

Let x be the original price. The amount of the mark-up is 15% of x, which is expressed .15x. The final price will be the original plus the markup, or **x + .15x**

Practice problem 7: Write an algebraic expression for the price of a pair of shoes with a discount of 20%.

Example 8: Write an algebraic expression for the amount of alcohol in a 35% solution.

Solution:

Let x be the original volume. The amount of alcohol 35% of x, which is expressed **.35x**.

Practice problem 8: Write an algebraic expression for the amount of acid in a 10% solution.

Example 9: Write an algebraic expression for the amount of alcohol in (x + 4) liters of a 20% solution.

Solution:

Let x + 4 be the original volume. The amount of alcohol 20% of (x+ 4) gallons, which is expressed **.20(x + 4)**.

Practice problem 9: Write an algebraic expression for the amount of acid (x − 3) gallons of a 15% solution.

Answers to practice problems:
1. 8n + 6
2. 7(n + 3)
3. 9n + 11

4. n= number of cats,
   56 − n = number of dogs
5. n = 1st integer
   n + 1 = 2nd integer
   n + 2 = 3rd integer

6. n = 1st even integer
   n + 2 = 2nd even integer
   n + 4 = 3rd even integer
7. x − .20x or .8x

8. .10x or .1x
9. .15(x − 3)

Exercises – Chapter 1

Write the following word phrases as algebraic expressions.

1. The sum of three times a number and six

2. The sum of twice times a number and seven

3. The difference of four times a number and three

4. The difference of twice a number and eight

5. Twice the difference of a number and nine

6. Five times the sum of a number and eleven

7. Half the sum of a number and six

8. The square of a number decreased by seven

9. Four times the square of a number increased by three

10. Three more than half a number

11. The value of x nickels and y dimes

12. The value of x dimes and y quarters

13. Four consecutive even integers

14. Four consecutive odd integers

15. John's and George's weight if George weighs 15 pounds more than John

16. The length of two pieces of a log which was 3 ft before cut.

17. The length and width of a pool whose length is twice the width.

18. Mary's age now, and Mary's age in three years

19. John's age now, and John's age four years ago

20. The amount of money in checking and saving accounts if the total money in both is $2,500

21. Three times a number is added to itself

22. Four times the difference of a number and three increased by five.

23. Seven more than twice a number added to its square

24. The cost of x items at $2 each

25. A salary of x dollars with a 15% raise

26. A price of x dollars with a discount of 20%

27. The amount of interest earned at 5% in one year on y dollars.

28. The cost of x gallans of gas at $3.39 per gallon

29. The amount of alcohol in x grams of a 25% solution

30. The amount of antifreeze in 25 – x liters of a 40% antifreeze solution.

31. The price of an oven with an original price of x dollars with a 20% mark-up

32. The price of a pair of jeans with a 50% discount

33. The cost of renting a car at $20 per day for an unknown time

34. The amount of acid in each solution if their total amount is 3 liters and the concentration is 20% for one and 35% for the other.

35. The time spent on two different jobs if the total time on both was 8 hours.

36. The value of 20 coins, all dimes and nickles

37. The value of 10 coins, all quarters and dimes

38. The amount of time on a round trip if the return trip only took seven eights as long

39. The distance traveled in x hours at 40 miles per hour.

40. The interest earned by x dollars invested for a year at 3%

Chapter 2 – Number Problems, Age Problems, Geometry Problems

Examples 1: Seven times the sum of a number and eight is eighty-four. Write an equation and find the number.
Solution:

First represent the unknown with a variable
x = the number

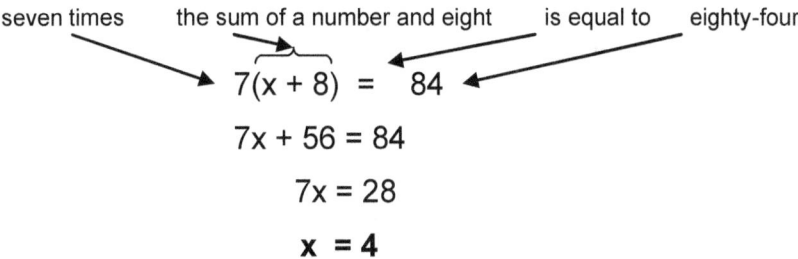

$$7(x + 8) = 84$$
$$7x + 56 = 84$$
$$7x = 28$$
$$\mathbf{x = 4}$$

Practice problem 1: Four times the sum of a number and nine is sixty-eight. Write an equation and find the number.

Example 2: The sum of three consecutive odd integers is 45. Find the integers.

Solution:
First represent the unknowns with algebraic expressions.
$\quad\quad x$ = the first consecutive odd integer
$x + 2$ = the second
$x + 4$ = the third

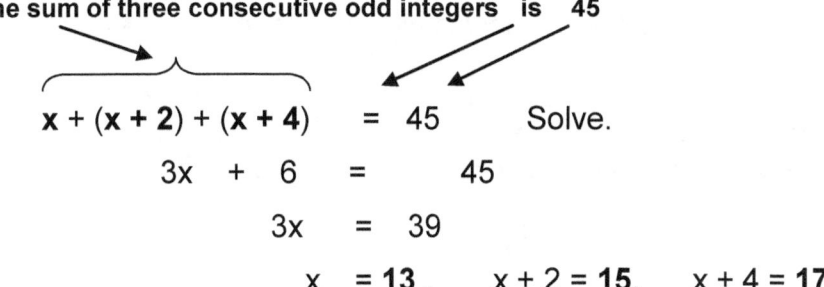

$$x + (x + 2) + (x + 4) = 45 \quad \text{Solve.}$$
$$3x + 6 = 45$$
$$3x = 39$$
$$x = 13, \quad x + 2 = 15, \quad x + 4 = 17$$

Practice problem 2: The sum of three consecutive odd integers is 33. Find the integers.

Example 3: John is three times as old as his brother. In four years he will be twice his brother's age. Find their ages.

Solution:

First represent the unknowns with algebraic expressions.
x = John's brother's age
3x = John's age

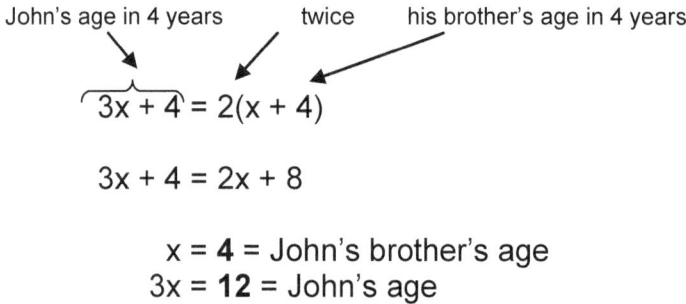

$$3x + 4 = 2x + 8$$

$$x = 4 = \text{John's brother's age}$$
$$3x = 12 = \text{John's age}$$

John is **12** years old, and his brother is **4** years old.

Practice problem 3: Mary is four times as old as Margaret. In six years, she will be only twice as old as Margaret. Find their ages.

Example 4: Two years ago, Eileen was half as old as her sister. In four years she will be three-fourths her sister's age. How old is Eileen now?

Solution:
First represent the unknowns with algebraic expressions.

	Eileen	Her sister
2 years ago	x	2x
4 years from now	x + 6	2x + 6

$$x + 6 = \frac{3}{4}(2x + 6)$$

$$4(x + 6) = 3(2x + 6)$$

$$4x + 24 = 6x + 18$$
$$-2x = -6$$
$$x = 3 = \text{Eileen's age 2 years ago}$$
$$2x = 6 = \text{Eileen's sister's age 2 years ago}$$
$$3 + 2 = \mathbf{5} = \text{Eileen's age now}$$

Practice problem 4: Three years ago Isaac was half his cousin's age. In twelve years he will be four-fifths his cousin's age. How old is Isaac now?

Example 5: One side of a triangle is three times another. The third side is 4 inches shorter than the longer side. If the perimeter is 31 inches, find the length of the sides.

Solution:
First represent the unknowns with algebraic expressions.

$x = 1^{st}$ side
$3x = 2^{nd}$ side
$3x - 4 = 3^{rd}$ side

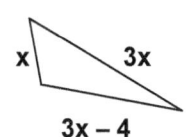

The perimeter is the sum of the sides.

$$x + 3x + (3x - 4) = 31 \quad \text{Solve for } x$$
$$x + 3x + 3x - 4 = 31$$
$$7x - 4 = 31$$
$$7x = 35$$
$$x = \mathbf{5} = \text{length of } 1^{st} \text{ side}$$
$$3x = \mathbf{15} = \text{length of } 2^{nd} \text{ side}$$
$$3x - 4 = \mathbf{11} = \text{length of } 3^{rd} \text{ side}$$

Length of sides are, **5** in, **15** in, and **11**

Practice problem 5: One side of a triangle is twice another. The third side is 6 inches shorter than the longer side. If the perimeter is 39 inches, find the length of the sides.

Answers to practice problems:

1. x = 8   **2.** 9, 11, 13   **3.** 3 yrs, Margaret; 12 yrs, Mary

4. 8 yrs.   **5.** 9, 18, 12

Exercises – Chapter 2

Write an equation and solve:

1. Eight more than twice a number is 26. Find the number.

2. Four less than three times a number is thirty-two. Find the number

3. If three times a number is added to the number the result is the number plus eighteen. Find the number

4. Four times a number minus six is three more than the number. Find the number.

5. Five times the difference of a number and seven is five more than the number. Find the number

6. Eleven more than four times a number is equal to one less than ten times the number. Find the number.

7. The sum of three consecutive integers is fifteen. Find the integers.

8. The sum of two consecutive even integers is twenty-six. Find the integers.

9. If three times the first consecutive integer is added to twice the second the result is twenty-seven. Find the integers.

10. If four times the first consecutive odd integer is added to the second, the result is one more than four times the second. Find the integers.

11. One number is 4 times a second number. If their sum is 80 what are the numbers?

12.   One-half a number is 2 more than one-third of the number. Find the number

13.   Two-thirds of a number is 3 more than half the number. Find the number.

14.   The difference of one-third a number and one-fourth the same number is 3. Find the number.

15.   The sum of two numbers is 34. Five times the smaller is 10 more than 3 times the larger number. Find the numbers.

16.   The difference of two numbers is 5. Three times the larger number is one more than 5 times the smaller number. Find the numbers.

17.   The sum of three numbers is 44. The second is 4 less than the first, and the third is twice the first. Find the numbers.

18.   The second number is 3 times the first and the third is twice the smaller. The sum of three numbers is 78. Find the numbers.

19.   One number is 7 more than another. Their sum is 29. Find the numbers.

20.   One number is 40 less than another. Their sum is 280. Find the numbers.

21.   Mary's mother's age is four times Mary's age. If Mary's mother was twenty-four when Mary was born, how old are each of them now?

22.   In eight years, Carrie will be twice as old as she was 8 years ago. How old is she now?

23.   Sixteen years ago, Mike's age was twice what it was twenty years ago. What is his current age?

24. John is twice as old as his brother. Six years ago he was four times as old as his brother. What are their ages now?

25. Adam is nine years older than George. In two years he will be four times as old as George. How old are each of them now?

26. Martha's mother is three times as old as Martha. In eight years, Martha will be the same age as her mother was sixteen years ago. How old are Martha and her mother now?

27. Daniel's mother's age is four times Daniel's age. In twelve years Daniel will be the same age his mother was fifteen years ago. Find their current ages.

28. Jacob is four years older than Terry. In five years the sum of their ages will be thirty-eight. How old is each of them now?

29. Isaac is six years older than Ben. In eight years the sum of their ages will be fifty-four. Find their current ages.

30. John is one-third as old as his father. Ten years from now he will be one-half as old as his father. How old is John now?

31. Joanne is half as old as her sister. In seven years she will be two-thirds as old as her sister. How old is her sister now?

32. Three years ago, Jerry was one sixth as old as his father. In eleven years he will be two-fifths as old as his father. How old is he now?

33. Joan is 8 years younger than Fran. Ten years ago, Joan was four-fifths as old as Fran. How old is each of them now?

34. Stephanie is six years younger than Daniel. In five years Stephanie will be five-sixths as old as Daniel. How old is each of them now?

35. Two years ago Iris was one-ninth as old as her mother. In sixteen years she will be seven-fifteenths as old as her mother. How old is Iris' mother now?

36. One side of a triangle is twice another. The third side is 8 inches less than the longer side. If the perimeter is 52, what are the lengths of each side?

37. One side of a triangle is three times another. The third is 10 less than twice the smaller. The perimeter is 110 inches. Find the length of the sides.

38. One side of a triangle 6 feet less than another. The third side is twice the smaller. The perimeter is 38. Find the length of the sides.

39. One side of a triangle is 9 feet less than another. The third is two thirds the larger. The perimeter is 63 meters. Find the length of the sides.

40. The width of a rectangle is six feet less than the length. The perimeter of the rectangle is 96 feet. Find the length and width.

41. The length of a rectangle is eight feet more than its width. The perimeter is 60 feet. Find the length and the width.

42. The length of a rectangle is four feet more than twice its width. If the perimeter is 146 feet, what are the length and width?

43. The length of a rectangle is three times its width. If the perimeter is 256 feet, what are the length and width?

44. The length of a rectangle is ten inches more than twice its width. If the perimeter is 170 feet, what are the dimensions of the rectangle?

45. One angle of a triangle is 10° less than twice the first. The second is 20° less than three times the first. Find the angles. (Hint: The sum of the angles of a triangle is 180°)

46. The second angle of a triangle is 6° less than twice the first. The third is 10° less than the first. Find the angles.

47. One angle of a triangle is twice another. The third is three-fourths the larger. Find the angles.

48. One angle of a triangle is one-third the second. The third is two-thirds the second. Find the angles.

49. One angle of a triangle is three-fourths the second. The third is half the second. Find the angles.

50. The second angle of a triangle is 12° more than half the first. The third is one and a half times the first. Find the angles.

Chapter 3 – Money Problems, Mixture Problems and Interest Problems

Example 1: Johnnie has 5 more nickels than dimes in his piggy bank, and his nickels and dimes together total $3.25. How many nickels and how many dimes are there?

Solution:

Let x = the number of dimes
x + 5 = the number of nickels

Write an equation (showing value in dollars).

$$.10x + .05(x + 5) = 3.25$$

where .10x is the value of dimes (number of dimes × value of dimes), .05(x + 5) is the value of nickels (number of nickels × value of nickels), and 3.25 is the total.

Solve.
$$.10x + .05x + .25 = 3.25$$
$$.15x = 3.00$$
$$x = \mathbf{20} \text{ dimes}$$
$$x + 5 = \mathbf{25} \text{ nickels}$$

Check: $.10(20) + .05(25) = 3.25$

$$2.00 + 1.25 = 3.25 \quad \textit{true}$$

So **20** dimes and **25** nickels is correct.

Practice problem 1: Jack has 6 more quarters than dimes in his piggy bank, and his dimes and quarters together total $5.00. How many quarters and how many dimes are there?

Example 2: Elizabeth has twice as many nickels as dimes. The rest of her 47 coins are quarters. The total value is $6.25. How many of each type of coin does she have?

Solution:

$$x = \text{the number of dimes}$$
$$2x = \text{the number of nickels}$$
$$47 - (x + 2x) = 47 - 3x = \text{the number of quarters.}$$

Write an equation (showing value in cents).

$$\underbrace{10x}_{\text{value of dimes}} + \underbrace{5(2x)}_{\substack{\text{value of nickels} \\ \times \text{ number of nickels}}} + \underbrace{25(47 - 3x)}_{\substack{\text{value of quarters} \\ \times \text{ number of quarters}}} = \underbrace{625}_{\text{total value}}$$

Solve. $10x + 5(2x) + 25(47 - 3x) = 625$

$10x + 10x + 1175 - 75x = 625$

$-55x = -550$

$x = \mathbf{10}$ dimes

$2x = \mathbf{20}$ nickels

$47 - 3x = \mathbf{17}$ quarters

Check: $10(10) + 5(20) + 25(17) = 625$

$100 + 100 + 425 = 625$ *true*

So **10**, **20** nickels and **17** quarters is correct.

Practice problem 2: Erin has twice as many dimes as nickels and 36 total coins. The rest are quarters. The total value of the coins is $6.50. How many of each type of coin are there?

A **solution** is a mixture containing more than one substance. For example, a 20% solution of acid contains 20% acid and 80% water. Pure water is 100% water and 0% acid. If a mixture of 30% acid and 50% acid is mixed it will result in a solution of a percentage between 30% and 50%. The percent of the solution is referred to as the "**concentration**"

Example 3: A chemist needs a 9 liters of 15% acid. He only has 5% solution and 20% solution. How much of each must he mix to obtain 15% solution.

Solution:

Let x = the volume of 5% solution
9 − x = the volume of 20% solution

Write an equation.

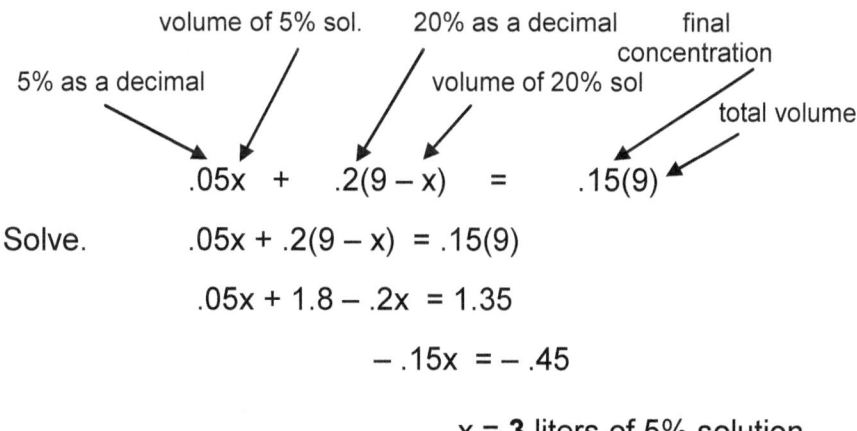

Solve.   .05x + .2(9 − x) = .15(9)

.05x + 1.8 − .2x = 1.35

− .15x = − .45

x = **3** liters of 5% solution
9 − x = **6** liters of 20% solution

Check: .05(3) + .2(6) = .15(9)
.15 + 1.2 = 1.35   true

Practice problem 3: A chemist needs a 40 liters of 20% acid. He only has 15% solution and 35% solution. How much of each must he mix to obtain 20% solution.

Example 4: How many ounces of a 25% salt solution must be added to 20 ounces of pure water to make a 5% solution?

Solution:

Let x = the volume of 25% solution
20 + x = the volume of 5% solution

Write an equation.

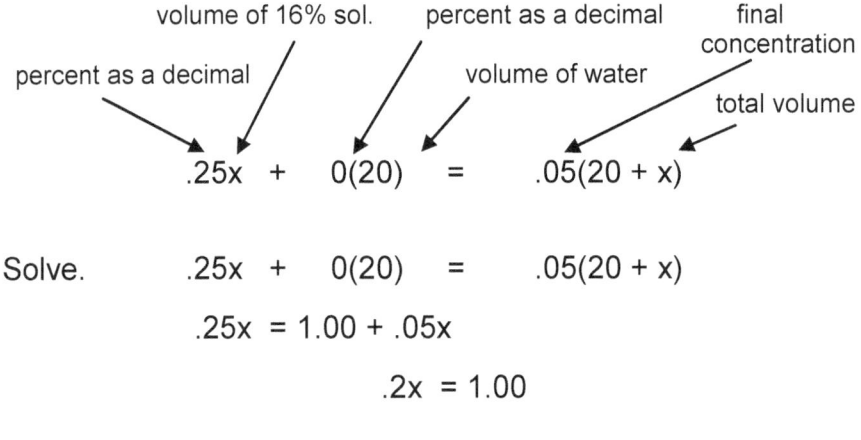

Solve.   .25x + 0(20) = .05(20 + x)

.25x = 1.00 + .05x

.2x = 1.00

x = **5** liters of 25% solution

Check: .25(5) + 0(20) = .05(25)
1.25 = 1.25)   *true*

Practice problem 4: How many ounces of water must be added to 6 ounces of an 8% solution of salt to make a 5% salt solution?

Example 5: An investor put part of $20,000 in an account paying 5% interest, and the remainder in an account paying 6%. If the total interest earned was $1080, how much did he invest in each.

Solution:

Let x = the amount at 5% interest
$20,000 − x = the amount at 6% interest

Write an equation.

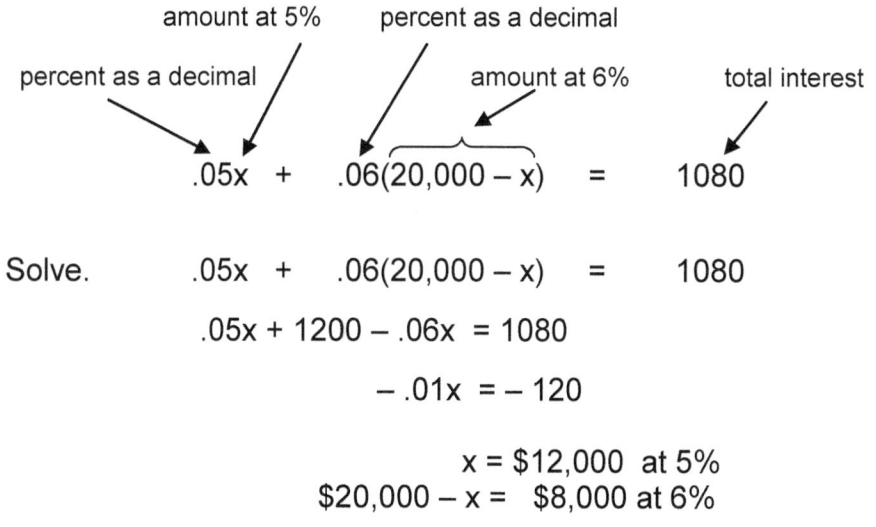

Solve.   .05x + .06(20,000 − x) = 1080

.05x + 1200 − .06x = 1080

− .01x = − 120

x = $12,000 at 5%
$20,000 − x = $8,000 at 6%

Check: .05(12,000) + .06(8000) = 1080
600 + 480 = 1080   true

Practice problem 5: John invested $10,000 in two accounts. One paid 9% and the other paid 12%. If the total interest earned was $1,110, how much was invested at each rate?

Answers to practice problems:

1. 10 dimes, 16 quarters
2. 5 nickels, 10 dimes, 21 quarters
3. 10 liters of 35%, 30 liters of 15%
4. 3.6 ounces of water
5. $3,000 at 9%, $7,000 at 12%

Exercises – Chapter 3

1. Jack has $4.00 in nickels and quarters. There are 32 coins. How many are nickels and how many are quarters?

2. Jane has 47 nickels and dimes. The total value is $3.40. How many each of nickels and dimes does she have?

3. Daniel has 40 dimes and quarters. They are worth $7.60. How many dimes and how many quarters does she have?

4. Juanita has six more quarters than dimes. The total value of her coins is $9.20. How many of each kind does she have?

5. Jerome has $3.10 in nickels and dimes. If he has eight more nickels than dimes, how many of each does he have?

6. Piper has 26 bills in $1, $5, and $10 denominations. Their total value is $99. She has twice as many ones as fives. How many of each denomination does she have?

7. Rylan has 92 coins worth $13. He has twice as many dimes as nickels, and the rest are quarters. How many coins of each kind does he have?

8. Mary's day job pays $8.50 per hour and her evening job pays $10.20 per hour. She worked 3 more hours at her evening job this week. If she made $311.10, how many hours did she work at each job?

9. Michele makes $10.50 per hour at one of her part-time jobs, and $7.75 at another. She works 5 fewer hours at the first job than at the second. If she made $312.50 in one week total at both jobs, how many hours did she work at each?

10. Jerry bought 100 pounds of nuts at at $2.63 a pound. If the mixture was made from two kinds, one worth $2.39 a pound and one worth $3.19 a pound, how much of each kind was used?

11. A tobacco salesman wants to make a mixture to sell at $3.95 a pound. If he uses two mixtures, one which sells for $4.75 a pound and one for $3.25 a pound, how much of each should he use to make 300 pounds?

12. A nursery wants to make 240 pounds of grass seed to sell for $1.17 a pound. How many pounds of $1.39 seed and how many pounds of $.84 seed should be mixed?

13. How many pounds of coffee worth $7 a pound should be mixed with 27 pounds of coffee worth $5.75 a pound to make a mixture to sell for $6.25 a pound?

14. How many pounds of coffee worth $8 a pound should be mixed with 24 pounds of coffee worth $4.50 a pound to make a mixture to sell for $5.90 a pound?

15. Isaac has $20 in dimes, quarters and half-dollars. He has 2 fewer dimes than six times the number of half-dollars, and there are 110 coins total. How many coins of each kind does he have?

16. Eugene has $7 in nickels, dimes and quarters. He has 5 more quarters than twice the number of dimes, and there are 39 total coins. How many are there of each kind?

17. There were 800 tickets sold at a concert. The total sales amounted to $4150. The ticket prices were $6.50 for the main floor and $4.50 for balcony. How many of each kind were sold?

18. How many ounces of a 30% salt solution must be added to 10 ounces of a 16% salt solution to make a 20% salt solution?

19. How many liters of a 16% acid solution should be added to 60 liters of a 3% acid solution to make an 8% acid solution?

20. How many quarts of water should be added to 2 quarts of a 10% solution of alcohol and water to make a 4% alcohol solution?

21. How much 60% alcohol solution should be added to 20 ounces of 45% alcohol to make a 55% alcohol solution?

22. How many grams of a 90% silver alloy should be melted with 40 grams of a 30% silver alloy to obtain a 50% silver alloy?

23. How many grams of a 70% gold alloy should be melted with 60 grams of a 20% gold alloy to obtain a 30% silver alloy?

24. How many gallons of 10% alcohol wine should be mixed with 40 gallons of 25% alcohol wine to produce a wine that is 15% alcohol?

25. How many pounds of pure salt should be mixed with 5 pounds of 40% salt solution to produce a 50% salt solution?

26. How many quarts of pure antifreeze should be mixed with 60 quarts of 45% antifreeze to obtain a 50% antifreeze solution?

27. How many ounces of a 4% peroxide solution should be added to 20 ounces of 30% solution to make a 12% solution?

28. How many quarts of an 80% alcohol solution should be added to 15 quarts of 6% alcohol solution to make a 20% solution?

29. If 100 ounces of a 90% silver alloy is mixed with 150 ounces of a 60% silver alloy, what is the percentage of silver in the final alloy?

30. If 20 grams of a 70% gold alloy is mixed with 55 grams of a 40% gold alloy, what is the percent of gold in the final mixture?

31. If 800 ounces of a 6% acid solution is mixed with 700 ounces of a 9% acid solution, what is the result of the mixture?

32. When 45 gallons of an 18% disinfectant solution is mixed with 60 gallons of a 32% disinfectant solution, what is the percent of disinfectant in the final mixture?

33. When 60 kilograms of a 30% copper alloy was mixed with 140 kilograms of a second alloy, the result was 65% copper. What was the percent of copper in the second alloy?

34. When 200 grams of 30% salt solution was mixed with 500 grams of a second solution, the result was a 20% salt solution. What was the percent of salt in the second solution?

35. When 30 ounces of a 46% iodine solution was mixed with 55 ounces of a second solution, the result was a 24% solution. What was the percent of iodine in the second solution?

36. A 40% copper alloy was mixed with a 90% copper alloy and the result was a 75% copper alloy. If there were 20 ounces more of the 90% alloy than the 40% alloy, now many ounces were there in the total mixture?

37. If a fertilizer that is 20% nitrogen is mixed with another fertilizer that is 60% nitrogen, the result is a fertilizer which is 34% nitrogen. There are 36 less of the 60% than the 20% fertilizer. How many pounds are there of the total mixture?

38. A mixture of 10,000 ounces of gold alloy with 4,000 ounces of a second alloy which is 35% less gold than the first. If the final alloy is 85% gold, what is the percentage of gold in each alloy?

39. If $30,000 is split into two accounts, one paying 8% and the other paying 12% interest per year, the interest earned totals $3,120. How much is invested in each account?

40. If $45,000 is split into two accounts, one paying 9% and the other paying 11% interest per year, the interest earned totals $4,630. How much is invested in each account?

41. If $9,000 is invested at 5.5%, how much needs to be invested at 12% for the interest from both investments to total $6,015?

42. If $10,000 is invested at 6%, how much needs to be invested at 7.5% for the interest from both investments to total $2,400?

43. The difference of two investments is $18,000. The smaller investment is 10% and the larger is 12%. The total income earned after one year is $3,920. How much was each investment?

44. Two equal amounts were invested, one at 5.25% and the other at 7.75%. The total income was $1040. What was the amount of the investments?

45. Isaac deposited part of his savings in an account paying 7% interest and three times as much in an account paying 11%. The total interest at the end of the year was $120. How much did he deposit in each account?

46. Monica invested part of her income tax refund at 6% and four times that amount at 11%. If the two investments earned $1000 at the end of the year, how much did she invest in each account?

47. Mary Ann wants to invest $1320 more in a CD paying 9% than she invests in one paying 12%. If the total interest from each is the same, how much will she invest at each rate?

48. Adam invests $1,260 more at 8% than he invests at 12%. The interest earned after a year is the same for both investments. How much did he invest in each account?

49. If the amount of interest earned by $15,000 is $75 less than the interest earned by $18,000 at .5% less per year, what is the rate of interest on each amount?

50. If the amount of interest earned by $8,000 is $150 less than the interest earned by $12,000 at .75% less per year, what is the rate of interest on each amount?

Chapter 4 – Motion Problems

We use the formula $d = rt$
Where; d is distance
       r is rate (speed)
       t is time

Example 1: John leaves for college driving 35 mph and leaves his book behind. His sister leaves half an hour later to take it to him driving 50 mph. How long does it take her to catch up with him?

Solution: t = the time John travels
                $t - .5$ = time his sister travels   ½ hr is .5

distance John travels = distance sister travels

rate x time (John) = rate x time (sister)

$$45t = 60(t - .5)$$

$$45t = 60t - 30$$

$$-15t = -30$$

$$t = \textbf{2 hours}$$

Practice problem 1: A ball team leaves for an away game in the bus travelling 50 mph, and the coach follows in his car 15 minutes later travelling 60 mph how long does it take him to pass them? (15 min = .25 hr)

Example 2: Two friends start driving toward each other from 280 miles away. One of them drives 10 mph faster than the other. If they meet in 2 hours, find the speed of each.

Solution: r = speed of slower car
r + 10 = speed of faster car

distance of slower car + distance of faster car = 280
rate x time of slower car + rate x time of faster car = 280

$$r(2) + (r + 10)(2) = 280$$

$$2r + 2r + 40 = 280$$

$$4r = 240$$

r = **60 mph** (speed of slower car)

r + 10 = **70 mph** (speed of faster car)

Practice problem 2: Jack and George are going to meet at a restaurant part-way between the cities in which they live which are 55 miles apart. If John travels 10 mph faster than George and it takes them 1/2 hour to get to the restaurant, how fast is each of them driving?

Example 3: A brother and sister leave home at the same time travelling in opposite directions. The brother is going 35 mph and his sister is driving 40 mph. How soon will they be 200 miles apart?

Solution: Let t = time

distance brother travels + distance sister travels = 150

rate x time of brother + rate x time of sister = 150

$$35t + 40t = 100$$

$$75t = 150$$

t = 2 hrs

Practice problem 3: Two planes take off from the same airport travelling in opposite directions. One is flying 80mph and the other is flying 90 mph. How long will it be before they are 255 miles apart?

Example 4: Two cars leave at the same time, going in the same direction. One is travelling 50 mph and the other is travelling 65 mph. How long will it take before they are 15 miles apart?

Solution: Let t = time

distance of faster − distance of slower = 15 miles

rate x time of faster − rate x time of slower = 15

$$65t - 50t = 15$$

$$5t = 15$$

$$t = 3 \text{ hours}$$

Practice problem 4: Two cars leave at the same time going the same direction. One is going 45 mph and the other is going 55 mph. How long will it take before they are 5 miles apart?

Answers to practice problems:

1. 1 ½ hours   2. John − 60 mph, George − 50 mph

3. 1 ½ hours   4. ½ hour

Exercises – Chapter 4

1. A cyclist starts on a course at 15 mph. Thirty minutes later, his friend starts from the same spot travelling 20 mph. How long does it take the second cyclist to overtake his friend?

2. A boat leaves the dock going 25 mph. An hour later another boat leaves the same dock going 35 mph. How long does it take the second boat to overtake the first?

3. A runner starts on a cross country course running an average of 5 mph. The second runner starts 15 minutes later running 6 mph. How long does it take the second runner to overtake the first?

4. John leaves his friend at the train depot and proceeds to the destination at 60 mph on a road which is parallel to the train tracks. His friend follows by train 45 minutes later, and the train travels 75 mph. How long does it take the train to overtake the car?

5. An airplane takes off from an airport flying 180 mph. A second airplane takes off in the same direction 30 minutes later travelling 240 mph. How long does it take to overtake the first plane?

6. Two busses start travelling toward each other from towns 300 miles away. If one bus averages 5 mph faster than the other one, and they meet in 4 hours, what is the speed of each bus?

7. Two cars travel toward each other from 160 miles apart. One is travelling 8 mph faster than the other, and they meet in 2 hours. What is the speed of each car?

8. Two runners start from opposite ends of a 15 mile cross country course. One runner runs 2 mph faster than the other. They meet after an hour and 15 minutes. How fast did each run?

9. Two trains travelled toward each other on parallel tracks from towns 290 miles apart. One train travel 5 mph faster than the other and they met after 2 hours. How fast did each train travel?

10. Two airplanes leave airports 650 miles away at the same time flying toward each other. One is flying 30 mph faster than the other. If they pass each other after two and a half hours, what is the speed of each plane?

11. Two runners start in opposite directions on a straight road. One averages at 5 mph and the other 7mph. How long will it be before they are 15 miles apart?

12. Two trucks leave the truck stop in opposite directions at the same, one at 55 mph and the other at 65 mph. In how many hours will they be 210 miles apart?

13. One airplane takes off travelling east at 160 mph, another takes off at the same time going west at 230 mph. How long will it take until they are 585 miles apart?

14. Two walkers start from the same place walking in opposite directions, one is walking 3 mph and the other is walking 2 mph. How long will it take them to get 7 miles apart?

15. Two trains start from the same station going opposite directions. One is travelling 70 mph, the other is going 80 mph. In how many hours will they be 525 miles apart?

16. Two hikers start from the same place on the same trail. One walks 3 mph, and the other walks 2 ½ mph. In how many hours will they be 3 miles apart?

17. Two trains start from the same station on parallel tracks going in the same direction. One is travelling at 80 mph and the other at 100 mph. How long will it be until they are 60 miles apart?

18. One runner finishes a race in 2 hours while the other runner is a quarter of a mile behind. If the first runner was running 3 miles an hour, what was the second runner's speed.

19. Graham left home at the same time as his brother and arrived at their grandparents in 3 hours. If Graham's speed was 75 mph, and his brother's speed was only 70 mph, how far away from his grandmother's house was his brother when Graham arrived?

20. How far apart did Graham and his grandparents live in problem (19) ?

Review

Write as an algebraic expression:

1. The sum of twice times a number and six

2. Twice the difference of a number and eight

3. John's age now, and John's age four years ago

4. Four times the difference of a number and three increased by five.

5. A salary of x dollars with a 15% raise

6. The amount of acid in each solution if their total amount is 3 liters and the concentration is 20% for one and 35% for the other.

Write an equation and solve:

7. Four less than three times a number is thirty-two. Find the number

8. The sum of three consecutive integers is fifteen. Find the integers.

9. One-half a number is 2 more than one-third of the number. Find the number

10. The sum of three numbers is 44. The second is 4 less than the first, and the third is twice the first. Find the numbers.

11. In eight years, Carrie will be twice as old as she was eight years ago. How old is she now?

12. Adam is nine years older than George. In two years he will be four times as old as George. How old are each of them now?

13. One side of a triangle 6 feet less than another. The third side is twice the smaller. The perimeter is 38. Find the length of the sides.

14. The length of a rectangle is seven inches less than twice its width. If the perimeter of the rectangle is 58 feet, what are the dimensions of the rectangle?

15. How many ounces of water must be added to 6 ounces of an 8% solution of salt to make a 5% salt solution?

16. Joan has 26 bills in $1, $5, and $10 denominations. Their total value is $99. She has twice as many ones as fives. How many of each denomination does she have?

17. Ann's day job pays $8.50 per hour and her evening job pays $10.20 per hour. She worked 3 more hours at her evening job this week. If she made $311.10, how many hours did she work at each job?

18. Alan bought 100 pounds of nuts at at $2.63 a pound. If the mixture was made from two kinds, one worth $2.39 a pound and one worth $3.19 a pound, how much of each kind was used?

19. How many ounces of a 30% salt solution must be added to 10 ounces of a 16% salt solution to make a 20% salt solution?

20. If $30,000 is split into two accounts, one paying 8% and the other paying 12% interest per year, the interest earned totals $3,120. How much is invested in each account?

21. A boat leaves the dock going 25 mph. An hour later another boat leaves the same dock going 35 mph. How long does it take the second boat to overtake the first?\

22. An airplane takes off from an airport flying 180 mph. A second airplane takes off in the same direction 30 minutes later travelling 240 mph. How long does it take to overtake the first plane?

23. Two runners start from opposite ends of a 15 mile cross country course. One runner runs 2 mph faster than the other. They meet after an hour and 15 minutes. How fast did each run?

24. Two trucks leave the truck stop in opposite directions at the same, one at 55 mph and the other at 65 mph. In how many hours will they be 210 miles apart?

25. One runner finishes a race in 2 hours while the other runner is a quarter of a mile behind. If the first runner was running 3 miles an hour, what was the second runner's speed?

Test

Write as an algebraic expression:

1. The sum of twice times a number and seven

2. Five times the sum of a number and eleven

3. The value of x nickels and y dimes

4. Seven more than twice a number added to its square

5. The amount of interest earned at 5% in one year on y dollars.

6. The value of 12 coins, all quarters and dimes

Write an equation and solve:

7. Four times a number minus six is three more than the number. Find the number.

8. If three times the first consecutive integer is added to twice the second the result is twenty-seven. Find the integers.

9. Two-thirds of a number is 3 more than half the number. Find the number.

10. Adam is nine years older than George. In two years he will be four times as old as George. How old are each of them now?

11. The length of a rectangle is ten inches more than twice its width. If the perimeter is 170 feet, what are the dimensions of the rectangle?

12. Jane has 47 nickels and dimes. The total value is $3.40. How many each of nickels and dimes does she have?

13. Rylan has 92 coins worth $13. He has twice as many dimes as nickels, and the rest are quarters. How many coins of each kind does he have?

14. There were 800 tickets sold at a concert. The total sales amounted to $4150. The ticket prices were $6.50 for the main floor and $4.50 for balcony. How many of each kind were sold?

15. Two hikers start from the same place on the same trail. One walks 3 mph, and the other walks 2 ½ mph. In how many hours will they be 3 miles apart?

Answers to Exercises

Chapter 1 (p. 19)

1. $3x + 6$
2. $2x + 7$
3. $4x - 3$
4. $2x - 8$
5. $2(x - 9)$
6. $5(x + 11)$
7. $\frac{1}{2}(x + 6)$ or $\frac{x + 6}{2}$
8. $x^2 - 7$
9. $4(x + 3)^2$
10. $\frac{1}{2}x + 3$

11. $5x + 10y$ cents
12. $10x + 25y$ cents
13. $x, x+2, x+4, x+6$
14. $x, x+2, x+4, x+6$
15. x = John's weight
    $x + 15$ = George's weight
16. x = lgth. 1st piece
    $3 - x$ = lgth. of 2nd piece
17. x = width, $2x$ = length
18. x = Mary's age now
    $x + 3$ = Mary's age in 3 yrs.
19. x = John's age now
    $x - 4$ = John's age 4 yrs. ago
20. x = checking, $2500 - x$ = savings

21. $3x + x = 4x$
22. $4(x - 3) + 5$
23. $x^2 + 2x + 7$
24. $2x$
25. $x + .15x$ or $1.15x$
26. $x - .2x$ or $.8x$
27. $.05y$
28. $3.39x$
29. $.25x$ gms.
30. $.40(25 - x)$ liters
31. $x + .20x$ or $1.2x$
32. $x - .50x = .5x$

33. $20x$
34. $.20x$ = acid in 20% sol
    $.35(3 - x)$ = acid in 35% sol
35. x = 1st job, $8 - x$ = 2nd job
36. x = # of dimes,
    $20 - x$ = # of nickels
    $10x + 5(20 - x)$
37. x = # of quarters,
    $10 - x$ = # of dimes
    $25x + 10(10 - x)$
38. $x + \frac{7}{8}x = 1\frac{7}{8}x$
39. $40x$
40. $.03x$

Chapter 2 (p. 17)

1. n = the number

    $2n + 8 = 26$
    n = **9**

2. $3n - 4 = 32$
    n = **12**

3. $n + 3n = n + 18$
    n = **6**

4. $4n - 6 = n + 3$
    n = **3**

5. $5(n - 7) = n + 5$
    n = **10**

6. $4n + 11 = 10n - 1$
    n = **2**

7. x = 1st
   x + 1 = 2nd
   x + 2 = 3rd

    $x + (x + 1) + (x + 2) = 15$

    4, 5, 6

8. x = 1st
   x + 2 = 2nd

    $x + (x + 2) = 26$

    12, 14

9. x = 1st
   x + 1 = 2nd

    $3x + 2(x + 1) = 27$

    5, 6

10. x = 1st odd integer
    x + 2 = 2nd odd integer

    $4x + (x + 2) = 4(x + 2) + 1$

    7, 9

11. 4x = the 1st
    x = the 2nd

    $4x + x = 80$

    16, 64

12. $\frac{1}{2}n = \frac{1}{3}n + 2$

    n = **12**

13. $\frac{2}{3}n = \frac{1}{2}n + 3$

    n = **18**

14. $\frac{1}{3}n - \frac{1}{4}n = 3$

    n = **36**

15. x = smaller
    34 - x = larger

    $5x = 3(34 - x) + 1$

    14, 20

16. x = 1st
    x - 5 = 2nd

    $3x = 5(x - 5) + 1$

    12, 7

17. x = 1st
$x - 4$ = 2nd
$2x$ = 3rd

$x + (x - 4) + 2x = 44$

12, 8, 24

18. x = 1st
$3x$ = 2nd
$2x$ = 3rd

$x + 2x + 3x = 78$

13, 39, 26

19. x = 1st

$x + 7$ = 2nd

$x + (x + 7) = 29$

11, 18

20. x = 1st
$x - 40$ 2nd

$x + (x - 40) = 280$

160, 120

21. x = Mary's age
$4x$ = Mary's mother's age

$24 + x = 4(24 + x)$

Mary **8 yrs**, mother **32 yrs**.

22. x = Carrie's age

$x + 8 = 2(x - 8)$

24

23. x = Mike's age now

$x - 16 = 2(x - 20)$

24

24. x = brother's age
$2x$ = John's age

$2x - 6 = 4(x - 6)$

John **18**, brother **9**

25. x = George's age
$x + 9$ = Adam's age

$(x + 9) + 2 = 4(x + 2)$

Adam **10**, George **1**

26. x = Martha's age
$3x$ = Mother's age

$x + 8 = 3x - 16$

Martha **12**, mother **36**

27. x = Daniel

$4x$ = mother

$x + 12 = 4x - 15$

Daniel **9**, mother **36**

28. x = Terry's age

$x + 4$ = Jacob's age

$(x + 5) + (x + 4 + 5) = 38$

Terry **12**, Jacob **16**

29. x = Ben's age

$x + 6$ = Isaac's age

$(x + 8) + (x + 6 + 8) = 54$

Ben **16**, Isaac **22**

30. x = father

$\frac{1}{3}x$ = John

$\frac{1}{3}x + 10 = \frac{1}{2}(x + 10)$

John **10**, father **30**

31. **x = sister**

$\frac{1}{2}x$ = Joanne

$\frac{1}{2}x + 7 = \frac{2}{3}(x + 7)$

Joanne **7**, sister **14**

32.   Jerry   Father

3 yrs ago $\frac{1}{6}x$ x

in 11 yrs $\frac{1}{6}x + 14$ $x + 14$

$\frac{1}{6}x + 14 = \frac{2}{5}(x + 14)$

$\frac{1}{6}(x - 3) = 7$

Father **36**, Jerry **6**

33. x = Fran

   $x - 8$ = Joan

$(x - 8) - 10 = \frac{4}{5}(x - 10)$

Fran **50**, Joan **42**

34. x = Daniel

   $x - 6$ = Sephanie

$(x - 6) + 5 = \frac{5}{6}(x + 5)$

Stephanie **25**, Daniel **31**

35.   Iris   Mother

2 yrs ago $\frac{1}{9}x$ x

$\frac{1}{9}x + 18 = \frac{7}{15}(x + 18)$

Mother is **27**

36. x = 1st
   $2x$ = 2nd
   $2x - 8$ = 3rd side

$x + 2x + (2x - 8) = 52$

12 in, **24** in, **16** in

37. x = 1st
   $3x$ = 2nd
   $2x - 10$ = 3rd

$x + 3x + (2x - 10) = 110$

20 in, **60** in, **30** in

38. x 1st
   $x - 6$ = 2nd
   $2(x - 6)$ = 3rd

$x + (x - 6) + 2(x - 6) = 38$

14 ft, **8** ft, **16** ft

39. x = 1st
x − 9 = 2nd
$\frac{2}{3}$x = 3rd

x + (x − 9) + $\frac{2}{3}$x = 63

27 ft, **18** ft, **18** ft

40. W = L − 6

2L + 2(L − 6) = 96

21 ft, width **27** ft length

41. L = W + 8

2W + 2(W + 8) = 60

11 ft, length **19** ft. width

42. L = 2W + 4

2W + 2(2W + 4) = 146

23 ft, width **50** ft length

43. L = 3W

W + 3W = 256

32 ft, width **96** ft length

44. L = 2W + 10

2W + 2(2W + 10) = 170

25 ft, width **60** ft length

45. x = 1st angle
2x − 10° = 2nd angle
3x − 20° = 3rd angle

x + (2x − 10) + (3x − 20) = 180

35°, 60°, 85°

46. x = 1st
2x − 6 = 2nd
x − 10 = 3rd

x + (2x − 6) + (x − 10) = 180

49°, 92°, 39°

47. x = 1st
2x = 2nd
$\frac{3}{4}$(2x) = 3rd

x + 2x + $\frac{3}{2}$x = 180

40°, 80°, 60°

48. $\frac{1}{3}$x = 1st
x = 2nd
$\frac{2}{3}$x = 3rd

$\frac{1}{3}$x + x + $\frac{2}{3}$x = 180

90°, 30°, 60°

49. $\frac{3}{4}x$ = 1st

　　　x = 2nd

　　$\frac{1}{2}x$ = 3rd

$\frac{3}{4}x + x + \frac{1}{2}x = 180$

60°, 80°, 40°

50. 　　　x = 1st

$\frac{1}{2}x + 12$ = 2nd

$1\frac{1}{2}x$ = 3rd

$x + (\frac{1}{2}x + 12) + \frac{3}{2}x = 180$

56°, 40°, 84°

Chapter 3   (p. 28)

1. 　x = no. of nickles
   32 - x = no. of quarters

   .05x + .2(32 - x) = 4.00

   20 nickels, **12** quarters

2. 　x = no of dimes
   47 - x = no of nickles

   .10x + .05(4 - x) = 3.40

   26 nickels, **21** dimes

3. 　x = no. of dimes
   40 - x = no. of quarters

   .10x + .25(40 - x) = 7.60

   16 dimes, **24** quarters

4. 　　x = no. of dimes
   x + 6 = no. of quarters

   .10x + .25(x + 6) = 9.20

   22 dimes, **28** quarters

5. 　　x = no. of dimes
   x + 8 = no. of quarters

   .10x + .25(x + 8) = 3.20

   26 nickels, **18** dimes

6. 　　x = no. of $5
   2x = no. of $1
   26 - 3x = no. of $10

   5x + 1(2x) + 10(26 - 3x) = 99

   14 ones, **7** fives, **5** tens

7. 　　**x = no. of nickles**
   2x = no. of dimes
   92 - 3x = no. of quarters

   .05x + .10(2z) + .25(92 - 3x) = 13.00

   20 nickels, **40** dimes,
   32 quarters

8. 　x = hrs at 8.50
   x + 3 = hrs. at 10.20

   8.50x + 10.20(x + 3) = 311.10

   15 hrs @ 8.50/hr
   18 hrs @ 10.20/hr

9. x = hrs @ 10.50/hr
    x - 5 = hrs @ 7.75/hr

10.50x + 7.75(x - 5) = 312.50

15 hrs @ $10.50,
20 hrs @ $7.75

10. x = lbs @ 2.39/lb
     100 - x = lbs @ 3.19/lb

2.39x + 3.19(100 - x) = 2.63(100)

70 lbs @ $2.39,
30 lbs @ $3.19

11. x = lbs. @ 4.75/lb
     300 - x = lbs. @ 3.25/lb

4.75x + 3.25(300 - x) = 3.95(300)

140 lbs @ $4.75
160 lbs @ $3.25

12. **x = lbs** @ 1.39/lb
     240 - x = lbs. @ .84/lb

1.39x + .84(240 - x) = 1.17(240)

144 lbs @ $1.29
96 lbs @ $ .84

13. x = lbs @ 7.00/lb

7.00x + 5.75(27) = 6.25(x + 27)

18 lbs.

14. **x = lbs.** @ 8.00/lb

8.00x + 4.50(24) = 5.90(x + 24)

16 lbs.

15. x = no. of half dollars
     6x - 2 = no. of dimes
     110 - (7x - 2) = no of quarters

.50x + .10(6x - 2) + .25(112 - 7x) = 20.00

70 dimes, **28** quarters,
12 half dollars

16. x = no. of dimes
     2x + 5 = no. of quarters
     39 - (3x + 5) = no. of nickles

.10x + .25(2x + 5) + .05(34 - 3x) = 7.00

7 nickels, **9** dimes,
23 quarters

17. x = no. for main floor
     800 - x = no. for balcony

6.50x + 4.50(800 - x) = 4150.00

275 main floor,
525 balcony

18. x = oz. of 30% sol.

.30x + .16(10) = .20(x + 10)

4 oz.

19. x = ltrs of 16% sol.

.16x + .03(60) = .08(x + 60)

37.5 liters

20. x = qts of water

0x + .10(2) = .04(x + 2)

3 qts.

21. x = oz. of 60%
.60x + .45(20) = .55(x + 20)
40 oz

22. x = gms. of 90%
.90x + .30(40) = .50(x + 40)
20 grams

23. x = gms of 70%
.70x + .20(60) = .30(x + 60)
15 grams

24. x = gal of 10%
.10x + .25(40) = .15(x + 40)
80 gal.

25. x = lbs of pure
1.00x + .40(5) = .50(x + 5)
1 lb.

26. x = qts of pure
1.00(x) + .45(60) = .50(x + 60)
6 qts.

27. x = oz of 4%
.04x + .30(20) = .12(x + 20)
45 oz.

28. x = qts. of 80%
.80x + ..06(15) = .20(x + 15)
3.5 qts.

29. x = % of silver final
.90(100) + .60(150) = x(100+150)
.72 = **72%**

30. x = % of gold final
.70(20) +.40(55) = x(20 + 55)
.48 = **48%**

31. x = % of final mixture
.06(800) + .09(700) = x(800 + 700)
.074 = **7.4%**

32. x = % of final mixture
.18(45) + .32(60) = x(45 + 60)
.26 = **26%**

33. x = % of copper in 2nd alloy
.30(60) + x(140) = .65(60 + 140)
.80 = **80%**

34. x = % of salt in 2nd. sol.
.30(200) + x(500) = .20(200 + 500)
.16 = **16%**

35. x = % of iodine in 2nd sol.
.46(30) + x(55) = .24(30 + 55)
.12 = **12%**

36. x = oz of 40%
x + 20 = oz of 90%
2x + 20 = oz of total
.40x +.90(x + 20) = .75(2x + 20)
50 oz.

37. x = lbs of 20%
x - 36 = lbs of 60%
2x - 36 = lbs. of total
.20x + .60(x - 36) = .34(2x - 36)
120 lbs.

38. x = % of gold in 1st
x - 35% = % of gold in 2nd
x(10,000) + (x - .35)4,000
= .85(10,000 + 4,000)
10,000oz of **95%,**
4,000 oz of **60%**

39. x = amt. at 8%
30,000 - x = amt. at 12%
.08x + .12(30,000 - x) = 3,120
$12,000 @ 8%,
$18,000 @ 12%

40. x = amt. at 9%
45,000 - x = amt at 11%
.09x + .11(45,000 - x) = 4,630
$16,000 @ 9%,
$29,000 @ 11%

41. x = amt at 12%
.055(9,000) + .12x = 6,015
$46,000

42. x = amt at 7.5%
.06(10,000) + .075x = 2,400
$24,000

43. x = amt at 12%
x - 18,000 = amt at 10%
.12x + .10(x - 18,000) = 3,920
$8,000 @ 10%,
$26,000 @ 12%

44. x = amt. invested in each
.0525x + .0775x = 1040
$8,000 each

45. x = amt at 7%
3x = amt at 11%
.07x + .11(3x) = 120
$300 @ 7%. $900 @ 11%

46. x = amt at 6%
4x = amt at 11%
.06x + .11(4x) = 1,000
$2,000 @ 6%, $8,000 @ 11%

47. x = amt at 12%
x + 1,320 = amt at 9%
.12x = .09(x + 1,320)
$5,280 @ 9%,
$3,960 @ 12%

48. x = amt at 12%
x + 1,260 = amt at 8%
.12x = .08(x + 1,20)
$3780 @ 8%,
$2,520 @ 12%

49. x = rate of $15,000
x - .05 = rate of $18,000
x(15,000) = (x - .05)18,000 - 75
$15,000 @ 5.5%,
$18,000 @ 5%

50. x = rate of $8,000
x - .75 = rate of $12,000
x(8,000) = (x - .75)12,000 - 150
$8, 000 @ **6%**
$12,000 @ **5.25%**

Chapter 4 (p. 39)

1. x = time of 2nd
   $x + .5$ = time of 1st
   $15(x + .5) = 20x$
   1½ hr

2. x = time of 2nd
   $x + 1$ = time of 1st
   $25(x + 1) = 35x$
   2½ hrs

3. x = time of 1st
   $5x = 6(x + 15)$
   1½ hrs

4. x = John's time
   $60x = 75(x - .25)$
   1 ¼ hr

5. x = time of 1st
   $180x = 240(x - .5)$
   2 hrs

6. x = speed of slow bus
   $4x + 4(x + 5) = 300$
   35 mph, 40 mph

7. x = speed of slow car
   $2x + 2(x + 8) = 160$
   36 mph, 44 mph

8. x = speed of slow runner
   $1.25x + 1.25(x + 2) = 15$
   5 mph, 7 mph

9. x = speed of slow train
   $2x + 2(x + 5) = 290$
   70 mph, 75 mph

10. x = speed of slow plane
    $2.5x + 2.5(x + 30) = 650$
    115 mph, 145 mph

11. t = time
    $5t + 7t = 15$
    1 hr, 15 min

12. t = time
    $55t + 65t = 210$
    1 hr, 45 min

13. t = time
    $160t + 230t = 585$
    1 ½ hrs

14. t = time
    $3t + 2t = 7$
    1 hr, 24 min

15. t = time
70t + 80t = 525
3 ½ hrs

16. t = time
3t - 2.5t = 3
6 hrs

17. t = time
100t - 80t = 60
3 hrs

18. t = time of 2nd
2(3) = 2t + .25
2.875 mph

19. x = distance
75(3) - 70(3) = x
15 mi.

20. x = distance
75(3) = x
225 mi

Review (p. 42)

1. $2n + 6$

2. $2(n - 8)$

3. x = John's age now
x − 4 = John's age 4 years ago

4. $4(n - 3) + 5$

5. x = salary now
x + .15x = **1.15x**
is salary with 15% increase

6. .20 x = acid in 20% sol.
.35(3 − x) = acid 35% sol.

7. 3n - 4 = 32
n = **12**

8. x = 1st
x + 1 = 2nd
x + 2 = 3rd
x + (x + 1) + (x + 2) = 15
the integers are **4, 5 and 6**

9. $\frac{1}{2}n = \frac{1}{3}n + 2$
n = **12**

10. x = 1st
x - 4 = 2nd
2x = 3rd
x + (x - 4) + 2x = 44
the numbers are **12, 8, 24**

11. x = Carrie's age now
x - 8 = Carrie's age in eight years
x + 8 = 2(x - 8)
x = **24** years old now

12. x = George's age
x + 9 = Adam's age
(x + 9) + 2 = 4(x + 2)
George- **1** yr Adam **-10** yrs

13. x = 1st side
    $x - 6$ = 2nd side
    $2x$ = 3rd side

    $x + (x - 6) + 2x = 38$

    11, 5, 22

14. $l = 2w - 7$

    $2l + 2w = P$

    $2(2w - 7) + 2w = 58$

    12 x 17

15. x = oz. of water

    $0x + .08(6) = .05(x + 6)$

    $x =$ **3.6** oz.

16. x = no of $5 bills
    $2x$ = no. of $1 bills
    $26 - 3x$ = no of $10 bills
    (26 minus the total number of $5 and $1)

    $5x + 1(2x) + 10(26 - 3x) = 99$

    $x =$ **7** fives
    $2x =$ **14** ones
    $26 - 3x =$ **5** tens

17. x = hrs. at $8.50/hr.
    $x + 3$ = hrs. at $10.20/hr.

    $8.50x + 10.20(x + 3) = 311.10$

    $x =$ **15** hrs at $8.50/hr.
    $x + 3 =$ **18** hrs at $10.20/hr.

18. x = lbs. at $2.39/lb.
    $100 - x$ = lbs. at $3.19/lb.

    $2.39x + 3.19(100 - x) = 2.63(100)$

    $x =$ **70** lbs. at $2.39/lb
    $100 - x =$ **30** lbs. at $3.19/lb.

19. x = oz. of 30% solution

    $.30x + .16(10) = .2(x + 10)$

    $x =$ **4** oz.

20. x = amt. at 8%
    $30,000 - x$ = amt. at 12%
    $.08x + .12(30,000 - x) = 3,120$
    $12,000 @ 8%,
    $18,000 @ 12%

21. x = time of 2nd
    $x + 1$ = time of 1st
    $25(x + 1) = 35x$
    2½ hrs

22. x = time of 1st
    $180x = 240(x - .5)$
    2 hrs

23. x = speed of slow runner
    $1.25x + 1.25(x + 2) = 15$
    5 mph, 7 mph

24. t = time
55t + 65t = 210
1 hr, 45 min

25. t = time of 2nd
2(3) = 2t + .25
2.875 mph

Test (p. 45)

1. 2n + 7

2. 5(n + 11)

3. 5x + 10y

4. $n^2 + (2x + 7)$

5. .05y

6. x = no of quarters
(12 − x) = no of dimes
25x + 10(12 - x)

7. 4n - 6 = n + 3
n = 3

8. n = 1st
n + 1 = 2nd

3n + 2(n + 1) = 27

1st = **5**   2nd = **6**

9. $\frac{2}{3}n = \frac{1}{2}n + 3$
n = 18

10. x = George's age
x + 9 = Adam's age

(x + 9) + 2 = 4(x + 2)

George - 1 yr.
Adam - 10 yrs.

11. L = 2W + 10

2W + 2(2W + 10) = 170

25 ft, width   **60** ft length

12. x = no of dimes
47 - x = no of nickles

.10x + .05(47 - x) = 3.40

26 nickels, **21** dimes

13. x = no. of nickles
2x = no. of dimes
92 - 3x = no. of quarters

.05x + .10(2z) + .25(92 - 3x) = 13.00

20 nickels, **40** dimes,
32 quarters

14. x = no. for main floor
800 - x = no. for balcony

6.50x + 4.50(800 - x) = 4150.00

275 main floor,
525 balcony

15. t = time
3t - 2.5t = 3
6 hrs

www.ingramcontent.com/pod-product-compliance
Lightning Source LLC
Chambersburg PA
CBHW061518180526
45171CB00001B/232